Beyond Compliance: Practical Cybersecurity for Schools

David Nanton

Published by David Nanton, 2025.

BEYOND COMPLIANCE: PRACTICAL CYBERSECURITY FOR SCHOOLS

First edition. March 6, 2025.

Copyright © 2025 David Nanton.

ISBN: 979-8230820079

Written by David Nanton.

Table of Contents

To my fellow soldiers in the IT trenches: those who battle printer jams, fight off phishing attacks, and wrestle Wi-Fi woes with equal valour.

May this book be a source of inspiration and perhaps a few much-needed laughs along the way.

Keep fighting the good fight!

INTRODUCTION
Real Actions for Real Protection

Imagine trying to herd a pack of hyenas. That's what it feels like sometimes, trying to manage cybersecurity in a school. They are the constant threats of data breaches, ransomware attacks, phishing scams, and now, even rogue AI trying to crash the party. They are sneaky and fast, and they are constantly evolving, looking for a way to pounce on our precious data and disrupt our digital lives.

The dangers are real, and the stakes are quite high. In 2024, the UK education sector faced a barrage of cyberattacks, with a worryingly high percentage of institutions reporting breaches. Government data [1] reveals that 71% of secondary schools, 86% of further education colleges, and a staggering 97% of higher education institutions experienced at least one cyberattack in the past year. To put that into perspective, only 50% of all UK businesses reported experiencing cyberattacks in the same period. This stark contrast underscores the heightened vulnerability of the education sector compared to other industries. This is further highlighted by the fact that 7% of educational institutions operate without any dedicated cybersecurity budget. The severity of these attacks is evident in real-world incidents that forced at least one school to temporarily close its doors after falling victim to a ransomware attack. [2]

But the UK is not alone in this. Educational institutions worldwide are grappling with similar cyber threats, with varying levels of preparedness and impact. According to Microsoft Threat Intelligence, the education sector is the third most targeted industry globally. [3] Moody's, a global credit rating provider, rated the education sector as having a "high" cyber risk in 2024[4], highlighting the sector's vulnerability and the potential for significant financial losses. These international trends underscore the global nature of cyber threats to education and the need for collaborative efforts to address them.

Why Practical Cybersecurity Matters

Schools hold a wealth of sensitive data, from pupil records and financial information to intellectual property and research. A successful cyberattack can not only disrupt operations, compromise sensitive information, and damage reputations, but also pose significant risks to the safety and wellbeing of the school community. And with the rise of sophisticated artificial intelligence (AI) technologies, the threats are becoming even more complex and challenging.

Indeed, AI is a double-edged sword. On one hand, it can be a powerful tool for enhancing security, automating threat detection, and bolstering defences. On the other, AI can be weaponised by attackers, used to create highly convincing phishing lures, generate sophisticated malware, and even manipulate individuals. The National Cyber Security Centre (NCSC) has warned that AI is likely to increase the scale and severity of ransomware attacks in the coming years.[5] Deepfake technology, which uses AI to create realistic fake videos and audio recordings, could be used in more sophisticated social engineering attacks, making it even harder to distinguish between legitimate and malicious communications.

Traditional cybersecurity measures, such as compliance exercises and certifications, while essential, are often insufficient in this dynamic and evolving landscape. A completed checklist or a certification achieved may provide a sense of accomplishment, but it can also create a false sense of security.

For instance, a school might be fully compliant with data protection regulations but still be vulnerable to phishing attacks due to a lack of staff training. Therefore, compliance should be seen as a baseline, not the end goal.

Many school IT departments also have cybersecurity policies and procedures in place, but these are ineffective if they are not actively enforced, regularly updated, and rigorously tested. Real-world threats require proactive actions, not just documentation.

A Practical Guide to Cybersecurity in Schools

This book aims to empower schools with the knowledge and tools they need to build a strong cybersecurity defence.

Now, you might be thinking: "This all sounds great, but our school's IT budget barely covers the cost of printer toners, let alone fancy AI-powered security tools!" And you'll be right – school budgets are often stretched thin and, with the best will in the world, cybersecurity might not always be at the top of the priority list. But the good news is that many of the key elements of a strong cybersecurity posture don't require a massive investment. Things like user education, strong passwords, and regular software updates can go a long way in protecting your school from cyber threats. Simply use this book to prioritise your efforts based on your school's specific needs and resources.

Throughout the book, I will be sharing some of my own adventures in cybersecurity – anecdotes from my 20-plus years in the trenches of school IT that illustrate the challenges and triumphs of navigating this ever-changing landscape. I have embellished some stories a bit for dramatic effect, but they are (mostly) true, and they offer valuable lessons learned from years of experience fighting those digital hyenas.

Beyond Compliance is your guide to building a cybersecurity culture where everyone in the school community understands their role in protecting the school's digital heart. It is about moving beyond a reactive, compliance-driven approach to embrace a proactive, hands-on strategy that can help your school thrive in the digital age, even with those dodgy hyenas lurking in the shadows.

1

THE HUMAN FIREWALL

User Education & Awareness

Cybersecurity in schools is not just about firewalls, intrusion detection systems, and the latest antivirus software. While these technological defences are crucial, they are rendered significantly less effective if the human element is neglected. Cybersecurity is fundamentally about people, and a robust security posture begins with a well-informed and vigilant school community. Engaging everyone – from pupils and staff to parents and governors – in cybersecurity awareness is paramount.

This chapter explores why user education and awareness must be the cornerstone of any school's cybersecurity strategy and how to implement a programme that cultivates a culture of security.

The Weakest Link: Human Error

I really like this statement by cybersecurity expert Kevin Mitnick, himself a former convicted hacker: "*Companies spend millions of dollars on firewalls and secure access devices, and it is money wasted because none of these measures address the weakest link in the security chain: the people who use, administer and operate computer systems.*"

This is particularly true in the educational environment. Phishing attacks, weak passwords, and careless online behaviour can compromise even the most secure systems, with the rise of sophisticated AI tools presenting new challenges. Therefore, building a "human firewall" through continuous education and awareness is the most effective and cost-efficient way to mitigate cyber threats.

The "Too-Good-to-be-True" Trap (or, How I Almost Fell for a Phishing Scam)

I like to think that I am fairly tech-savvy. But even I have had my embarrassing slip-ups. Not too long ago, I received an email that looked like it was from my mobile phone provider. It said there was some suspicious activity on my account, and I needed to click on a link to verify my information. Now, alarm bells should have been ringing right away, but I was distracted, on my mobile, and the email looked so official! I clicked the link, entered my details... and then it hit me. This was a classic phishing scam! Luckily, I realised my mistake quickly enough to change my password and also contact my bank, but it was a close call. It just goes to show that anyone can fall victim to these scams, no matter how tech-savvy they think they are. And with AI making these scams even more sophisticated, we need to be more vigilant than ever.

Regular Training: Embedding Cybersecurity in the School Culture

Occasional workshops or annual training sessions are insufficient for fostering a lasting impact on cybersecurity awareness. Schools must move beyond this reactive approach and integrate cybersecurity education into the fabric of the school culture. Regular, ongoing training for all staff and pupils is essential. This training should not be a one-size-fits-all approach. Content should be tailored to different age groups and roles within the school. For instance, training for younger pupils might focus on basic concepts like online safety and responsible sharing, while older pupils could delve into more complex topics like data privacy, the implications of cyberbullying, and the ethical use of AI. Staff training should cover areas such as data protection regulations, password management best practices, recognising and reporting suspicious activity, and understanding the potential risks and benefits of AI tools in education.

These training sessions should be engaging and interactive, moving beyond dry presentations. Consider incorporating real-world examples, case studies, and interactive exercises to reinforce key concepts. Regular refresher sessions are vital to keep cybersecurity at the forefront and to address emerging threats. Cybersecurity is a dynamic field, and new attack vectors are constantly being developed. Therefore, ongoing training ensures that the school community remains up to date and prepared for the latest threats.

Pupil Education: Cultivating Digital Citizens

Children today are often immersed in the digital world from a very young age. Schools have a responsibility to equip them with the knowledge and skills necessary to navigate this landscape safely and responsibly. Cybersecurity education should be integrated into the curriculum from prep school onwards with a focus on both traditional online safety and the unique challenges posed by AI. This should include age-appropriate lessons on topics such as:

- **Online safety:** Safe browsing habits, avoiding inappropriate content, and understanding the risks of online interactions.
- **Data privacy:** Protecting personal information, understanding the implications of sharing data online, and managing privacy settings.
- **Cyberbullying:** Recognising and responding to cyberbullying, understanding the impact of online harassment, and promoting respectful online communication.
- **Digital citizenship:** Developing responsible online behaviour, understanding copyright and intellectual property, and recognising the ethical implications of technology use.
- **AI Literacy:** Understanding how AI works, its potential benefits and risks, and how to interact with AI systems safely and responsibly. This includes recognising the limitations of AI and the potential for bias and misinformation.

By embedding these concepts in the curriculum, schools can cultivate a generation of digital citizens who prioritise cybersecurity and understand their role in protecting themselves and others online.

Phishing Simulations: Learning by Doing

Phishing attacks are one of the most common and effective methods used by cybercriminals. Simulating phishing attacks can be a powerful tool for raising awareness and identifying areas for improvement. These simulations involve sending realistic-looking phishing emails to staff and pupils to test their ability to recognise and report them. The results of these simulations can provide valuable insights into the effectiveness of any training and highlight areas where further education is needed. It is crucial to emphasise that these simulations are not intended to trick or embarrass individuals but rather to provide a safe learning environment where mistakes can be made without real-world consequences. Following a simulation, it is essential to provide feedback and reinforce best practices for identifying and reporting suspicious emails.

Parent Engagement: Extending Cybersecurity Beyond the School Gates

Cybersecurity awareness should not be confined to the school environment. Children and young people are exposed to online risks at home and elsewhere. Therefore, engaging parents in cybersecurity education is also important. Schools can provide parents with resources and information on how to keep their children safe online, including workshops, online resources, and information sessions on topics such as:

- **Parental controls:** Setting up parental controls on devices and online services to restrict access to inappropriate content and monitor online activity.
- **Social media safety:** Understanding the risks of social media, managing privacy settings, and promoting responsible online interactions.
- **Online gaming safety:** Understanding the risks of online gaming, protecting personal information, and promoting safe online interactions.
- **AI safety at home:** Educating parents about the potential risks of AI tools and applications used by children at home.

- **Open communication:** Encouraging open communication with children about their online experiences and fostering a culture of trust.

By working in partnership with parents, schools can create a consistent message about cybersecurity and ensure that children are protected both at school and at home. This collaborative approach is essential for building a truly secure and resilient school community.

In conclusion, user education and awareness are not merely components of an effective cybersecurity strategy; they are the foundation upon which all other security measures are built. By prioritising the human element, schools can create a network of vigilant individuals who are empowered to identify and report threats, ultimately strengthening the entire school's defence against cyberattacks.

The Neighbourhood Watch

Think of your school as a friendly neighbourhood. There are strong locks on your doors (firewalls), CCTV watching the streets (intrusion detection systems), and maybe even a guard dog (antivirus software). These are great for protecting your home, but what if your neighbours aren't careful? What if they leave their doors unlocked, fall for scams, or invite strangers into the neighbourhood? Suddenly, your own home is at risk, even with all your security measures.

User education is like creating a "Neighbourhood Watch" for your school. It's about teaching everyone in the neighbourhood – pupils, teachers, parents, even the lollipop lady – how to spot suspicious activity. It's about teaching them not to click on strange links in emails (don't talk to strangers!), how to create strong passwords (lock your doors!), and what to do if they see something suspicious (report it!).

2

GUARDING THE DIGITAL GATES

Strengthen Access Controls

Access control is a critical line of defence against unauthorised access to sensitive data and systems. It is about ensuring that only authorised individuals can access specific resources, limiting the potential damage from both internal and external threats. Effective access control is not simply about erecting a digital wall; it is about implementing a layered approach that combines technical safeguards with clear policies and procedures.

This chapter delves into the essential principles and practices of strengthening access controls within a school environment, focusing on the concepts of least privilege and multi-factor authentication.

Least Privilege: Allow Only What's Necessary

The principle of least privilege is a cornerstone of robust access control. It dictates that users should only be granted the minimum level of access necessary to perform their job functions. This means that a teacher would have access to department files and pupil records, but not necessarily to the school's financial data. Similarly, a pupil should have access to their own files and learning resources, but not to administrative systems. By limiting access on a need-to-know basis, the potential impact of a compromised account or malicious insider is significantly reduced. Even if an account is compromised, the damage is contained because the user's access was limited in the first place.

Implementing least privilege requires a thorough understanding of the different roles and responsibilities within the school. A comprehensive access control matrix should be developed, outlining which users or groups have access to which resources. This matrix should be regularly reviewed and updated. Regular audits are essential to identify and revoke privileges that are no longer required. For example, when a staff member leaves the school, their access should be immediately revoked. Similarly, if a pupil graduates or transfers, their access to school systems should be terminated.

AI adds a layer of complexity to least privilege. AI-powered systems may require access to multiple data sources to function effectively. It is important to carefully consider the level of access granted to these AI systems, ensuring they only have access to the data they absolutely need. Furthermore, as AI systems evolve and their roles change, their access privileges may also need to be adjusted dynamically. This requires a flexible and adaptable access control framework.

IOIO
IOIO

The Case of the "Borrowed" Login

I once knew a colleague with a bit of a shortcut habit. Whenever someone needed access to something on the school network, and they couldn't remember their password, he'd just lend them his. "It's just quicker this way," he'd say. "You can get your password reset later."

Well, one day, a colleague used his login to make some changes to the pupil database, and let's just say it didn't go as planned. Chaos ensued, and he learned his lesson the hard way: least privilege is not just a good idea, it is essential! Each person should only have access to what they absolutely need. No more "borrowing" logins!

Implementing least privilege can be challenging, especially in large schools with numerous users and systems. However, the benefits far outweigh the challenges. By limiting access, schools can reduce the risk of data breaches, protect sensitive information, and improve overall security posture.

Multi-Factor Authentication: Adding an Extra Layer of Security

Passwords on their own are no longer enough to protect sensitive systems and data. Even strong passwords can be cracked through phishing, brute-force attacks, or data breaches. Multi-factor authentication (MFA) adds an extra layer of security by requiring users to provide two or more verification factors to gain access to a system. These factors can be categorised as:

- **Something you know:** This is typically a password or PIN.
- **Something you have:** This could be a smartphone, a hardware token, or a smart card.
- **Something you are:** This refers to biometric authentication, such as a fingerprint, facial recognition, or iris scan.

MFA is not foolproof, but it significantly reduces the risk of unauthorised access, even if a password is compromised. An attacker would need to possess multiple factors to gain access, making it much more difficult to breach the system. For example, even if an attacker steals a user's password, they would still need access to the user's smartphone to complete the second factor of authentication.

Implementing MFA should be a priority for schools, especially for sensitive systems and privileged accounts. This includes accounts with access to pupil data, financial information, and administrative systems. While MFA can add a small amount of complexity to the login process, the added security is well worth the inconvenience. There are various MFA solutions available, from simple text message codes to more sophisticated biometric authentication systems. Schools should choose the solutions that best fit their needs and budget.

AI & Access Control: Opportunities and Challenges

AI presents both opportunities and challenges for access control. On the one hand, AI can be used to enhance access control by automating tasks, detecting anomalies, and providing real-time risk assessments. For example, AI-powered systems can analyse user behaviour to identify potential insider threats or compromised accounts. AI can also be used to dynamically adjust access privileges based on user context, such as location, time of day, or device used.

On the other hand, AI also poses new security risks. For example, AI can be used to bypass traditional MFA methods or impersonate users. Additionally, AI systems can themselves also be vulnerable to attacks, so if an AI system used for access control is compromised, it could potentially grant unauthorised access to sensitive data and systems.

Therefore, schools need to carefully consider the security implications of using AI in access control. It is crucial to ensure that AI systems are properly secured and that appropriate safeguards are in place to prevent misuse.

Combining Least Privilege and MFA for a Powerful Defence

The combination of least privilege and MFA provides a powerful defence against unauthorised access. By granting users only the necessary access and requiring multiple factors for authentication, schools can significantly reduce the risk of data breaches and other security incidents. These two principles work together to create a layered security approach that is much more difficult to penetrate than a single layer of defence.

In addition to implementing these technical safeguards, schools should also develop clear policies and procedures related to access control, including policies on password management, account creation and termination, and IT acceptable use. Regular training should be conducted to educate staff and pupils about these policies and the importance of access control.

Strengthening access controls requires continuous monitoring and improvement. Schools should regularly review their access control policies and procedures, conduct security audits, and stay up to date on the latest threats and vulnerabilities. By taking a proactive approach to access control, schools can protect sensitive data, maintain the integrity of their systems, and create a more secure learning environment for pupils and staff. In conclusion, robust access control, built upon the principles of least privilege and multi-factor authentication, forms a crucial pillar of any school's cybersecurity framework, safeguarding the digital gates and ensuring that sensitive information remains protected.

The VIP Room

Imagine your school has a "VIP Room" where all the really sensitive material is kept. Access control is like deciding who gets a key to this room and what they are allowed to do inside.

Least Privilege is like giving different people different keys. The Head gets a master key that opens everything. Teachers get keys that only open the cabinets with their pupils' marks. You only give people access to what they absolutely need to do their job.

MFA is like adding extra locks to the VIP room door. Maybe you need a key *and* a secret code to get in, so even if someone steals your key, they still can't get in without the code.

By combining least privilege and MFA, you make it much harder for anyone who shouldn't be in the VIP room to get in.

3

BUILDING A FOUNDATION OF RESILIENCE

Secure Devices & Software

A robust and secure technical infrastructure is the bedrock of effective cybersecurity in schools. Unfortunately, many schools adopt a reactive, patchwork approach to security, layering defences onto existing systems without considering the overall architecture. This often leads to vulnerabilities and gaps that can be exploited by cybercriminals.

This chapter advocates for a proactive approach, emphasising building security into the digital infrastructure from the outset, using key elements like network segmentation, regular patching, strong endpoint security, and secure configuration.

Network Segmentation: Dividing & Conquering Threats

Network segmentation is a fundamental security practice that involves dividing a network into smaller, isolated sub-networks or segments. This approach offers several advantages in strengthening a school's cybersecurity posture. By isolating different parts of the network, segmentation limits the "blast radius" of a cyberattack. If one segment is compromised, the impact is contained, preventing the malware or attacker from easily spreading to other critical areas of the network.

In a school environment, network segmentation can be used to separate various user groups and systems. For example, the administrative network, which houses sensitive pupil data and financial information, could be segmented from the pupil network. This would prevent a pupil's compromised device from providing an attacker with access to the administrative network. Similarly, the guest network, if one exists, should be isolated from the internal network to prevent unauthorised access to sensitive resources. Other potential segments could include a dedicated network for IoT (Internet of Things) devices, a separate network for staff devices, and a secure zone for critical servers.

Implementing network segmentation requires careful planning and design. It is crucial to understand the different types of traffic flowing across the network and to define clear boundaries between segments. Firewalls, virtual LANs (VLANs), and other network security tools can be used to enforce segmentation policies. Regular monitoring and testing are also essential to ensure that segmentation is effective, and that traffic is flowing as intended.

The Rogue Robot Rampage

It was a Friday afternoon (of course!), and I was just about to head out for the weekend when suddenly, chaos erupted in the school's robotics lab. One of the robots had gone haywire, spinning around the lab, bumping into things, knocking over equipment, and generally causing mayhem. A pupil had been tinkering with R2-D2's programming and accidentally unleashed its mischievous side. Thankfully, we had recently implemented network segmentation, separating the robotics lab network from the rest of the school's. This prevented R2 from accessing any sensitive data or causing widespread disruption. We managed to contain the rogue robot to the lab and restore order, but it was a close call. That incident demonstrated how network segmentation is not just about protecting against external threats; it's also about containing internal "threats," even those with a mischievous robotic twist. It's like having separate playpens for toddlers and teenagers – it keeps everyone safe and prevents unnecessary chaos.

Security Patches: Plugging the Holes

Software and operating systems are constantly being updated to fix bugs, improve performance, and address security vulnerabilities. Cybercriminals often target known vulnerabilities in outdated software to gain access to systems. Therefore, keeping software and systems up to date with the latest security patches is crucial for defending against cyber threats. Schools should establish a routine for regular updates and patches to ensure that all software, operating systems, and applications are protected against the latest threats.

Patch management can be a complex task, especially in a school environment with numerous devices and software applications. Manual patching can be time-consuming and prone to errors. Whenever possible, schools should automate the patching process. Many operating systems and software applications offer automatic update features that can be configured to install patches as soon as they are released. Centralised patch management systems can also be used to deploy patches to multiple devices simultaneously.

It is important to prioritise patching based on the severity of the vulnerability. Critical vulnerabilities that could allow an attacker to gain control of a system should be addressed immediately. Regular vulnerability scanning can help identify systems that are missing patches and prioritise patching efforts. Patching should not be limited to operating systems and applications. Firmware updates for network devices, servers, and other hardware components are also important for maintaining security.

Endpoint Security: Protecting the Front Lines

Endpoint devices, such as laptops, desktops, tablets, and smartphones, are often the first point of contact for cyberattacks. These devices are vulnerable to malware infections, phishing attacks, and other threats. Therefore, robust endpoint security is essential for protecting the entire school network. Schools should implement strong antivirus and endpoint detection and response (EDR) solutions on all devices.

Antivirus software scans files and programs for known malware signatures and prevents them from executing. EDR solutions go beyond traditional antivirus by monitoring endpoint activity for suspicious behaviour and providing tools for detecting and responding to advanced threats. EDR solutions can also help identify and isolate compromised devices, preventing the spread of malware to other parts of the network.

In addition to antivirus and EDR, other endpoint security measures include:

- **Firewall:** A personal firewall on each device can help block unauthorised access to the device.
- **Data loss prevention (DLP):** DLP software can prevent sensitive data from leaving the school network.
- **Device encryption:** Encrypting hard drives can protect data if a device is lost or stolen.
- **Mobile device management (MDM):** MDM solutions can help manage and secure mobile devices, such as smartphones and tablets.

Secure Configuration: Locking Down the System

Secure configuration involves configuring devices and software in a way that minimises security risks. This includes turning off unnecessary services and features, setting strong passwords, and implementing access controls. Many devices and software applications come with default configurations that are not secure. Schools should review the default settings and make changes to improve security.

For example, unused ports and services should be disabled to reduce the attack surface. Default passwords should be changed to strong, unique ones. Access controls should be implemented to restrict access to sensitive data and systems. Regular security assessments and penetration testing can help identify configuration weaknesses.

Secure configuration should be an ongoing process. As new threats and vulnerabilities are discovered, configurations may need to be adjusted. Schools should stay up to date on security best practices and implement changes as needed. By taking a proactive approach to securing devices and software, schools can create a more resilient digital infrastructure and protect themselves against cyber threats.

This layered approach, combining network segmentation, regular patching, strong endpoint security, and secure configuration, forms a formidable defence, allowing schools to focus on their core mission of teaching and learning, knowing their digital environment is fortified against the rapidly evolving landscape of cyber threats.

Your Digital Castle

Think of your school's network as a royal castle. You want to make sure it's strong and secure so no invaders (hackers) can get in and cause trouble.

Network Segmentation: This is like building walls within your castle, dividing it into different sections. You have the royal chambers (administrative network), the training grounds (pupil network), and the guest quarters (guest network). If the invaders manage to sneak into the guest quarters, they can't easily get to the royal chambers because of the walls. It limits the damage.

Security Patches: This is like regularly checking your castle walls for any cracks or weaknesses and patching them immediately. If you leave a crack open, the invaders will find it and exploit it! Keeping your software patched is a bit like keeping your castle walls in top shape.

Your Digital Castle (cont'd)

Endpoint Security: These are like having guards posted at every entrance to your castle, checking everyone who comes in and making sure they aren't carrying any nasty surprises (malware). They're also watching for anyone acting suspiciously.

Secure Configuration: This is like making sure all the doors and windows in your castle are locked properly. You wouldn't leave a window open in the royal treasury, would you? Secure configuration means turning off any unnecessary features that could be used by invaders and setting strong passwords for all the important rooms.

By building your digital castle with strong walls (network segmentation), regular maintenance (security patches), vigilant guards (endpoint security), and locked doors and windows (secure configuration), you create a much more resilient defence against any invaders trying to cause chaos.

4

PROTECTING THE FOUNDATION

Physical Security

While the focus of cybersecurity often centres on digital threats like malware and hacking, the importance of physical security cannot be overstated. A robust cybersecurity strategy must encompass both the digital and physical aspects, recognising that a weakness in either can compromise the entire system.

This chapter explores the essential elements of physical security in schools, emphasising access control, surveillance systems, and environmental controls. Think of your school's IT infrastructure as a valuable treasure chest. Cybersecurity protects the contents, but physical security protects the chest itself.

Access Control: Guarding the Inner Sanctum

Physical access control is a key element of school security. As with digital access control, it is about ensuring that only authorised individuals have access to sensitive areas, such as server rooms, network cabinets, and other sensitive locations. These areas may house critical infrastructure, including servers, network equipment, and backup systems or sensitive information such as exam papers. Restricting access minimises the risk of unauthorised individuals tampering with equipment, stealing data, or causing damage. Effective access control measures include:

- **Physical Barriers:** Implementing physical barriers, such as locked doors and security fences, can deter unauthorised access. Key card or biometric access systems can provide an additional layer of security, allowing only authorised staff to enter restricted areas.

- **MFA for Physical Access:** Just as MFA is crucial for digital access, it can also be implemented for physical access. Combining a key card with a PIN or biometric scan adds an extra layer of security, making it more difficult for unauthorised individuals to gain access.
- **Visitor Management:** Implementing a visitor management system can help track who is entering and exiting restricted areas. Visitors should be escorted by authorised staff and required to sign in and out.
- **Regular Audits:** Regularly audit access logs and review access control policies to ensure they are up to date and effective. Remove access privileges for individuals who no longer require it.
- **Clear Policies and Procedures:** Establish clear policies and procedures regarding access to restricted areas. Communicate these policies to all staff and ensure they are understood and followed.

Surveillance: Keeping a Watchful Eye

Surveillance systems, such as CCTV cameras and other monitoring devices, play a crucial role in deterring crime and monitoring sensitive areas. These systems act as a virtual security guard, providing a watchful eye over critical infrastructure and helping to identify suspicious activity. Surveillance systems can:

- **Deter Crime:** The presence of CCTV cameras can act as a deterrent to potential intruders. Knowing that they are being monitored can discourage individuals from attempting to gain unauthorised access.

- **Monitor Sensitive Areas:** CCTV cameras can be strategically placed to monitor server rooms and other critical areas, enabling schools to keep an eye on these locations and identify any suspicious activity.
- **Investigate Incidents:** CCTV footage can be invaluable in investigating security incidents. If a break-in or other incident occurs, the recordings can be used to identify the culprits and understand the sequence of events.
- **Provide Evidence:** CCTV footage can also be used as evidence in legal proceedings.

When implementing surveillance systems, schools should consider:

- **Camera Placement:** Cameras should be strategically placed to cover all critical areas, ensuring there are no blind spots.
- **Recording Quality:** Ensure that the cameras record high-quality footage that can be used for identification and investigation purposes.
- **Storage and Retention:** Establish clear policies for storing and retaining any footage and ensure that it is stored securely and retained for an appropriate period.
- **Privacy Considerations:** Be mindful of privacy considerations when implementing surveillance systems. Ensure that the systems are used in a way that respects the privacy of pupils, staff, and visitors.

The Great Bin Heist

I never thought I'd be using our CCTV system to solve a neighbourhood dispute, but hey, that's school life for you. It all started when Miss Stern, who also happened to live in one of the school residences, stormed in, brandishing a misplaced teacup and fuming about the disappearance of her bins. I must admit, my initial reaction was to stifle a laugh. Stolen bins? Really? But then I remembered our school's commitment to security, which extended to the wellbeing of our resident staff. So, the IT team put on our detective hats and got to work. We reviewed the CCTV footage, and lo and behold, spotted Miss Stern's neighbour, Mr. Grumbles, casually wheeling her bins into his own yard in the dead of night! Faced with the evidence (and Miss Stern), Mr. Grumbles rather sheepishly returned the bins. And Miss Stern brought us chocolates the next day! The Great Bin Heist might have been a minor incident, but it highlighted how our security measures could be used in unexpected ways to support the school community. Who knows, maybe IT will be mediating fence disputes next!

Environmental Controls: Protecting from the Elements

Protecting IT equipment from environmental threats is just as important as protecting it from the digital ones. Environmental factors, such as fire, water damage, and power outages, can cause significant damage to critical infrastructure and disrupt school operations. Implementing environmental controls can help mitigate these risks. Key considerations include:

- **Fire Suppression:** Install fire suppression systems in server rooms and other sensitive areas. Ensure that these systems are regularly inspected and maintained.
- **Water Leak Detection:** Install water leak detectors in server rooms to alert staff of any leaks. This can help prevent water damage to sensitive equipment.
- **Temperature and Humidity Control:** Maintain appropriate temperature and humidity levels in server rooms to prevent equipment from overheating or being damaged by excessive moisture. Use HVAC systems and environmental monitoring tools to ensure optimal conditions.
- **Power Backup**: Power is technology's Achilles heel. The most advanced systems will be pretty useless if you can't power them on! Implement uninterruptible power supplies (UPS) to provide backup power in the event of a power outage. This can help ensure that critical systems remain operational during power disruptions.
- **Physical Security against Natural Disasters:** In areas prone to natural disasters such as floods, ensure that server rooms are built to withstand such events. This may involve reinforcing walls, elevating equipment, or implementing other protective measures. Think of location too – basements

are never an ideal location for server rooms as water always flows downwards, so even an internal leak can be damaging.

Building a Comprehensive Physical Security Strategy

Implementing a comprehensive physical security strategy requires a holistic approach that includes:

- **Risk Assessment:** Conduct a thorough risk assessment to identify potential physical security threats.
- **Layered Security:** Implement multiple layers of security, combining physical barriers, surveillance systems, and environmental controls.
- **Regular Inspections:** Conduct regular inspections of physical security measures to ensure they are in good working order.
- **Training and Awareness:** Train staff on physical security procedures and raise awareness about the importance of physical security.
- **Policy and Procedures:** Develop clear policies and procedures related to physical security and communicate them to all staff.

By prioritising physical security, schools can create a more secure environment for their IT infrastructure, protecting valuable data and ensuring the continuity of critical services. It is a fundamental part of a comprehensive cybersecurity strategy, recognising that protecting the digital world starts with securing the physical foundation.

Fort Knox

Think of your school's server room as the school's "Fort Knox," holding all the valuable digital assets.

Access control is like having a strong gate and a well-trained guard at the entrance, ensuring that only authorised personnel can enter.

Surveillance systems are like having security cameras monitoring the perimeter, keeping a watchful eye on any suspicious activity.

Environmental controls are like having a climate-controlled vault inside Fort Knox, protecting the valuable contents from fire, water damage, and other environmental threats.

Just like the real Fort Knox, your school's server room needs to be protected from both internal and external threats. Physical security measures, such as access control, surveillance, and environmental controls, are essential for ensuring the safety and integrity of the school's digital assets.

5

SAFEGUARDING THE SCHOOL'S DIGITAL HEART

Data Protection, Business Continuity &
Disaster Recovery

From pupil records and academic performance to financial information and staff details, schools rely on data to function effectively. Protecting this data is not just a best practice; it is a legal and ethical imperative.

This chapter explores the crucial aspects of data protection, disaster recovery, and business continuity, emphasising the importance of encryption, maintaining a data asset register, implementing regular backups, and leveraging data loss prevention (DLP) strategies. A robust strategy in these areas ensures that even in the face of cyberattacks, natural disasters, or human error, the school can continue to operate and protect its valuable information.

Data Encryption: Locking Down Sensitive Information

Data encryption is a fundamental security measure that transforms readable data into an unreadable format, known as ciphertext. This ensures that even if the data is intercepted by unauthorised individuals, it cannot be understood without the decryption key. Schools should implement encryption for both data at rest (stored data) and data in transit (data being transmitted). This is especially critical for data processed by AI systems, as this data may be particularly sensitive or used to train AI models.

- **Data at Rest Encryption:** This protects data stored on hard drives, servers, USB drives, and other storage media. If a device is lost or stolen, the encrypted data remains secure. Full-disk encryption, which encrypts the entire hard drive, is a particularly effective method for protecting data at rest. Database encryption and file-level encryption are other options, depending on the specific needs and systems used by the school.

- **Data in Transit Encryption:** This protects data as it is being transmitted across a network, whether within the school or over the internet. Secure protocols like HTTPS, which uses SSL/TLS encryption, should be used for all web traffic. Virtual Private Networks (VPNs) can be used to create secure connections between the school network and remote locations. Email encryption can protect sensitive information sent via email.

Implementing encryption requires careful planning and execution. Schools should choose encryption solutions that are appropriate for their specific needs and ensure that encryption keys are securely managed. Key management is crucial, as the loss of an encryption key can leave the data permanently inaccessible.

The Data Asset Register: Knowing Your Data Landscape

You cannot protect your data if you don't know where it is. Maintaining a comprehensive data asset register is essential for understanding the school's data landscape. This register should document all software and platforms that store school data, including cloud services, on-premise servers, and even physical storage locations.

Critically, this register must also include data used and processed by AI systems. For each data asset, the register should include information on:

- **Data type:** What type of data is stored (e.g., pupil records, financial data, staff information, AI training data, AI model outputs)?
- **Location:** Where is the data stored (e.g., server name, cloud service provider, AI training data repository, AI model storage location)?
- **Data owner:** Who is responsible for the data?
- **Data retention requirements:** How long must the data be kept?
- **Contract dates:** When do contracts with vendors expire?
- **Security measures:** What security measures are in place to protect the data?

Maintaining a data asset register is an ongoing process. The register should be regularly reviewed and updated to reflect changes in the school's data landscape. Establishing clear IT procurement procedures can help, by ensuring staff are not acquiring or signing up to data systems without anyone's knowledge. This information is invaluable for incident response, data breach notification, and compliance with data privacy regulations.

Data Backups: The Cornerstone of Recovery

Regular backups are one of the most effective ways to prepare for a cyberattack, especially ransomware, or any data loss event. A robust business continuity and disaster recovery (BCDR) plan should include a strategy for regularly backing up critical data and systems. Backups should be performed frequently, ideally on a schedule that aligns with the importance of the data. Critical data, such as pupil records, should be backed up more frequently than less critical data.

Backups should be stored in a secure location, preferably offsite or in the cloud. This ensures that backups are not affected by a disaster or cyberattack that impacts the school's primary systems. It is crucial to regularly test the restoration process to ensure that backups can be restored quickly and reliably. Testing should simulate different scenarios, such as a server failure or a ransomware attack, to validate the BCDR plan and identify any weaknesses. Regular testing will help ensure that, in the event of a breach or disaster, the school can recover quickly with minimal data loss and disruption to operations.

The "Day the Internet Died" (or, Why Disaster Recovery is Not Just for Hollywood Movies)

It was a dark and stormy morning (okay, maybe it was just a typical Thursday), and I was enjoying my first cup of coffee when the phone rang. It was Mrs. Panicked. "The internet is down!" she shrieked, "Everything is down! We can't access anything!" My heart skipped a beat. This was not good. It took some investigation, but I eventually discovered the cause of the outage: a rogue squirrel had decided to take a nibble on our main internet cable. Yes, this is true, a squirrel! It was like something out of a cartoon, but the consequences were very real. The entire school was cut off from the digital world. No emails, no online resources, no access to the MIS. It was chaos.

But thankfully, I had a secret weapon: our recently updated disaster recovery plan. We quickly switched

The Day the Internet Died (cont'd)

over to our backup internet connection, restored essential services from our offsite backups, and within a couple of hours, we were back online.

The squirrel was apprehended (and hopefully given a stern lecture about the importance of cybersecurity), and the school day was saved.

That day, the school learned that disaster recovery is not just for Hollywood movies; it's a critical part of any school's cybersecurity strategy. It's like having a spare tire in your car – you hope you never need it, but you're incredibly grateful when you do.

Data Loss Prevention (DLP): Preventing Data Exfiltration

Data loss prevention (DLP) solutions help prevent sensitive data from leaving the school network without authorisation. DLP tools can monitor network traffic, endpoint devices, and cloud applications to detect and block attempts to exfiltrate data. DLP strategies can include:

- **Restricting access to removable media:** Preventing the use of USB drives and other removable media can help prevent data theft.
- **Blocking access to unauthorised file sharing sites:** Blocking access to file sharing sites not sanctioned by the school can prevent employees or pupils from uploading sensitive data to external platforms.
- **Monitoring email and web traffic:** DLP solutions can monitor email and web traffic for sensitive data and block or quarantine messages or files that contain such data.
- **Implementing data masking:** Data masking techniques can be used to hide sensitive data when it is not needed for a particular task, reducing the risk of accidental or intentional data exposure.

Implementing DLP calls for a thorough understanding of the types of data that need to be protected and the potential channels for data exfiltration. DLP policies should be carefully designed and implemented to minimise disruption to legitimate activities while effectively preventing data loss.

A Holistic Approach to Data Protection

Data protection, disaster recovery, and business continuity are interconnected elements of a comprehensive cybersecurity strategy. By implementing the measures outlined in this chapter – encryption, data asset register, regular backups, and DLP – schools can significantly strengthen their defences against data loss and ensure business continuity in the face of unforeseen events. This proactive approach to data management not only safeguards sensitive information but also demonstrates a commitment to responsible data stewardship, building trust within the school community and fulfilling legal and ethical obligations. Regular review and adaptation of these strategies are essential to keep pace with evolving threats and ensure the ongoing protection of the school's digital heart.

The Time Capsule

Think of your school's data as a precious time capsule filled with all sorts of important memories and information – pupil photos, reports, school play scripts, even the secret recipe for the catering team's famous pizza. You want to make sure this time capsule is safe and can be opened in the future, even if something bad happens.

Data Encryption: This is like locking the time capsule with a strong, unbreakable lock. Even if someone finds the time capsule, they can't open it without the key (decryption key). It keeps the information inside safe and secret.

Data Asset Register: This is like having a detailed inventory of everything inside the time capsule. You know exactly what's in there, who put it in, and when it should be opened. It helps you keep track of all the important stuff.

Regular Backups: This is like making copies of

The Time Capsule (cont'd)

everything in the time capsule and storing them in a safe place, maybe even a different location. If something happens to the original time capsule, you still have the backups to restore everything. It's like having a spare key and a map to where it's hidden!

Data Loss Prevention (DLP): This is like having a security system around the time capsule to prevent anyone from sneaking in and taking things out. Maybe you have a motion sensor or a guard dog to make sure no one messes with the precious contents.

By encrypting the data (locking the time capsule), keeping a data asset register (inventorying the contents), making regular backups (creating copies), and implementing DLP (having a security system), you ensure that your school's precious "time capsule" of data is safe, protected, and can be accessed when needed, even if disaster strikes.

6

PREPARING FOR THE INEVITABLE

Incident Response Planning

Cybersecurity certifications and compliance exercises are valuable tools for establishing a baseline of security practices. However, they can never fully prepare schools for the chaotic reality of an actual cyberattack. A well-defined and regularly practiced incident response plan is crucial for navigating the complexities of a cyber incident, minimising damage, and ensuring a swift and effective recovery.

This chapter explores the essential components of incident response planning, emphasising the importance of a comprehensive plan, regular practice, and clear communication.

Plan: Building a Robust Incident Response Framework

The foundation of any effective incident response strategy is a well-defined incident response plan. This document serves as a roadmap for how the school will handle a cyber incident, outlining roles and responsibilities, communication strategies, technical procedures, and the steps to be taken to contain, eradicate, and recover from an attack. The plan should be comprehensive, addressing various types of incidents, from ransomware attacks and data breaches to denial-of-service attacks and malware infections. Key elements of a robust incident response plan include:

- **Incident Response Team:** Identify a dedicated incident response team composed of individuals with diverse skills and expertise. This team should include representatives from

IT, senior leadership, administration, and other relevant departments. Clearly define the roles and responsibilities of each team member, ensuring that everyone understands their part in the response process. A designated incident response manager should be appointed to lead the team and coordinate activities.

- **Incident Classification:** Establish a clear system for classifying incidents based on their severity and impact. This allows the team to prioritise their efforts and allocate resources effectively. A common classification system uses levels or tiers to categorise incidents, ranging from minor security events to major incidents that require immediate action.

- **Containment Strategy:** Define procedures for containing an incident to prevent further damage. This may involve isolating affected systems, disabling compromised accounts, or blocking malicious traffic. The containment strategy should be tailored to the specific type of incident and should prioritise minimising disruption to critical services.

- **Eradication Procedures:** Outline the steps for eradicating the threat, which may involve removing malware, patching vulnerabilities, or rebuilding compromised systems. Eradication should be performed carefully to ensure that the threat is completely removed and does not resurface.

- **Recovery Process:** Detail the procedures for restoring systems and data to their pre-incident state. This should include restoring from backups, rebuilding servers, and reconfiguring network devices. The recovery process should be prioritised to restore critical services first.

- **Communication Plan:** Develop a clear communication plan to keep stakeholders informed during an incident. This plan should identify who needs to be notified, when they need to be notified, and how they will be notified. Stakeholders may include staff, pupils, parents, the police and regulatory bodies like the Information Commissioner's Office (ICO). The communication plan should also address how to handle any media queries and manage the school's reputation during an incident.

- **Legal Considerations:** Address any legal and regulatory requirements related to incident response, such as data breach notification laws. Consult with legal advisors to ensure that the incident response plan complies with all applicable laws and regulations.
- **Post-Incident Analysis:** Outline the process for conducting a post-incident analysis to identify the root cause of the incident, evaluate the effectiveness of the response, and develop recommendations for preventing future incidents. This analysis is crucial for learning from the incident and improving the school's overall security posture.

Practice: Rehearsing for Reality

Having a plan is only the first step. The plan must be regularly updated and tested to ensure its effectiveness and to familiarise the incident response team with their roles and responsibilities. Regular training and exercises are essential for preparing the team to respond effectively to a real incident. Two common types of exercises are:

- **Tabletop Exercises:** These are facilitated discussions where team members walk through a simulated incident scenario, discussing their actions and responsibilities. Tabletop exercises are a valuable way to test the plan and identify any gaps or weaknesses.

- **Mock Incidents:** These are more realistic simulations that involve actively responding to a simulated incident. Mock incidents can be used to test the team's technical skills and their ability to work together under pressure.

Regular practice is crucial for ensuring that the incident response team is prepared to handle a real incident. It helps to identify areas where the plan needs to be improved and ensures that everyone is familiar with their role in the process.

The Great Ransomware Panic

It all started with a seemingly harmless email that Mr. Click-Happy, our resident staff tech enthusiast (and, unfortunately, most frequent clicker of suspicious links), opened without a second thought. Bam! Ransomware. Suddenly, files were encrypted, and a menacing message popped up on our screens, demanding a hefty sum of Bitcoin to release our precious data. Panic swept through the school. Teachers were frantic, admin staff were bewildered, and even the Head looked a bit pale. But we had only recently practiced our incident response plan so it was quite fresh in our memories. We quickly activated our BCDR procedures and, within 24 hours, we were back up and running. Mr. Click-Happy got a very stern talking-to (and a mandatory refresher course on cybersecurity awareness), and the school learned a valuable lesson about the importance of having a well-rehearsed incident response plan. It's like learning to ride a bike – you might fall a few times in practice, but when you really need to ride, you'll be glad you put in the time to learn.

Communicate: Keeping Stakeholders Informed

Effective communication is essential during a cyber incident. Keeping stakeholders informed helps to manage expectations, build trust, and minimise the impact of the incident. The communication plan should be activated as soon as an incident is suspected. Key communication considerations include:

- **Timeliness:** Communicate with stakeholders promptly and regularly. Provide updates as the situation evolves and avoid delays in communication.
- **Transparency:** Be transparent about the incident and its potential impact. Avoid downplaying the severity of the incident or withholding information.
- **Accuracy:** Ensure that all communication is accurate and factual. Avoid speculation or rumours.
- **Consistency:** Maintain consistent messaging across all communication channels.
- **Channels:** Use a variety of communication channels to reach different stakeholders, such as email, instant messaging, phone calls, and the school website.
- **Target Audience:** Tailor communication to the specific audience. Different stakeholders may require different levels of detail and information.
- **Designated Spokesperson:** If required, designate a spokesperson to communicate with the media and the public. This ensures consistent messaging and prevents conflicting information from being released.

By prioritising planning, practice, and communication, schools can develop a robust incident response capability that minimises the impact of cyber incidents and ensures a swift and effective recovery. This proactive approach to incident response is essential for protecting the school's data, reputation, and operational continuity in the face of ongoing cyber threats. It moves beyond simply reacting to attacks and instead builds a culture of preparedness, ensuring the school is ready to face the inevitable challenges of the digital age.

The School's Fire Drill

What if your school has a fire? What do you do? You don't just run around screaming, right? You have a fire drill! You have a plan (incident response plan), you practice it regularly (practice), and you make sure everyone knows what to do (communicate). A cyberattack is like a digital fire, and you need a digital fire drill – an incident response plan – to deal with it effectively.

Plan: This is like drawing up the fire escape map and assigning roles – who's in charge of evacuating which classroom, who calls the fire services, etc. Your incident response plan is your digital fire escape map, telling everyone what to do in case of a cyberattack.

Practice: This is like actually practising the fire drill. You walk through the steps, make sure everyone knows where to go, and time how long it takes to evacuate. Practicing your incident response plan through tabletop exercises and mock incidents is like practicing your digital fire drill. It helps you find any

The School's Fire Drill (cont'd)

kinks in the plan before a real "fire" breaks out.

Communicate: This is like making sure everyone knows about the fire drill and where to go. Your communication plan is how you keep everyone informed during a cyberattack – staff, pupils, parents, even the media. You need to know who to contact, when, and how.

Just like a well-practiced fire drill can save lives in a real fire, a well-defined and practiced incident response plan can save your school from serious damage during a cyberattack. It's about being prepared for the inevitable, so you can react quickly and effectively when disaster strikes.

7

THE SILENT GUARDIAN

Network Monitoring

In the complex digital anatomy of a modern school, the network acts as the central nervous system, connecting everything from pupil devices and administrative systems to critical infrastructure. Protecting this vital network requires constant vigilance and proactive monitoring.

This chapter delves into the critical aspects of network monitoring, focusing on Intrusion Detection and Prevention Systems (IDPS) and Security Information and Event Management (SIEM) solutions, emphasising their role in building a robust defence-in-depth strategy. Think of your school network as a bustling city. Network monitoring is like having a dedicated security team constantly patrolling the streets, keeping an eye out for suspicious activity, and ready to respond to any trouble.

Intrusion Detection and Prevention Systems: The First Line of Defence

Intrusion Detection and Prevention Systems (IDPS) act as the first line of defence against malicious network activity. These systems continuously monitor network traffic for suspicious patterns and known attack signatures. Think of IDPS as the CCTV and motion sensors of your school's network. They are constantly watching for anything out of the ordinary. An IDPS can be deployed in various ways, including network-based IDPS, which monitors traffic across the entire network, and host-based IDPS, which monitors activity on individual devices.

IDPS solutions can perform several key functions:

- **Intrusion Detection:** IDPS can detect suspicious activity, such as port scanning, denial-of-service attacks, and malware infections. It will then generate an alert, notifying your IT team.
- **Intrusion Prevention:** Some IDPS solutions can actively block or prevent malicious activity. For example, an IDPS can block traffic from known malicious IP addresses or terminate connections that are exhibiting suspicious behaviour.
- **Traffic Analysis:** IDPS can analyse network traffic to identify trends and anomalies. This can help IT departments understand network usage patterns and identify potential security risks.
- **Log Management:** IDPS can generate logs of network activity, which can be used for security analysis and incident response.

Deploying IDPS is a crucial step in building a defence-in-depth strategy. This strategy recognises that no single security tool is foolproof. It assumes that an attacker may be able to bypass one layer of security, so multiple layers of defence are implemented to increase the overall security posture. IDPS plays a vital role in this strategy by providing an early warning system for potential attacks.

Security Information and Event Management (SIEM): The Central Intelligence Hub

While IDPS solutions focus on real-time monitoring and prevention, Security Information and Event Management (SIEM) systems provide a more holistic view of network security. SIEM solutions collect and analyse security logs from a wide range of sources, including IDPS, firewalls, servers, and applications. Think of a SIEM as the central intelligence hub where all the security information from across the school network comes together. It's like having a team of analysts who can piece together all the clues to get a complete picture of what's happening.

SIEM solutions can help schools:

- **Identify and Respond to Threats:** By correlating security events from multiple sources, SIEM can identify complex attacks that might go unnoticed by individual security tools. For example, a series of seemingly unrelated events, such as a failed login attempt followed by unusual file access, could indicate a potential intrusion.
- **Improve Incident Response:** SIEM can provide valuable information for incident response investigations. By analysing security logs, security teams can understand the timeline of events, identify the scope of the attack, and determine the root cause.
- **Meet Compliance Requirements:** Many regulatory frameworks require organisations to maintain security logs and monitor for security events. SIEM solutions can help schools meet these compliance requirements.
- **Gain Visibility into Network Activity:** SIEM provides a centralised view of network activity, allowing security teams to monitor network usage patterns, identify potential security risks, and improve overall security posture.

How SIEM Saved the School Play

It was a week before the school play, and tensions were running high. Costumes were being misplaced, lines were being forgotten, and the music teacher was threatening to quit if someone didn't find her missing baton. Then, disaster struck: the internet ground to a halt, rehearsals were disrupted, and the entire production was on the verge of collapse. Our SIEM system quickly pointed me in the right direction: the drama teacher had decided to upload a massive video file of the dress rehearsal, completely hogging the bandwidth. Apparently, she wanted to share the "sneak peek" with parents but hadn't quite grasped the concept of file compression. We had a good laugh about it, she learned a valuable lesson about internet etiquette, and the play went on to be a huge success. The incident highlighted the power of SIEM in not only detecting anomalies but also providing valuable insights into network usage patterns. It's like having a traffic report for your network, showing where the congestion is and helping things flow along smoothly.

Building a Comprehensive Network Monitoring Strategy

Implementing effective network monitoring requires careful planning and execution. Schools should:

- **Identify Critical Assets:** Determine which systems and data are most critical and prioritise monitoring efforts accordingly.
- **Choose the Right Tools:** Select IDPS and SIEM solutions that meet the budget and specific needs of the school.
- **Configure Tools Properly:** Ensure that the tools are configured correctly to monitor the appropriate traffic and generate relevant alerts.
- **Establish Alerting Procedures:** Develop clear procedures for responding to security alerts.
- **Regularly Review and Update:** Network monitoring tools and strategies should be regularly reviewed and updated to keep pace with evolving threats.
- **Train Staff:** Ensure that relevant IT staff are trained on how to use the monitoring tools and respond to security alerts.

Network monitoring is not a one-time project. By continuously monitoring the network and analysing security logs, schools can identify and respond to threats more quickly and effectively, minimising the impact of cyberattacks and ensuring the continued operation of critical systems. Just as the security team in our analogy is always on guard, network monitoring tools, perhaps augmented by AI capabilities, provide a constant, silent guardian for the school's digital infrastructure.

The Security Team

Imagine your school as a large, complex building. The network is like the intricate system of pipes, wires, and ventilation ducts that keep the building running. IDPS are like the security guards patrolling the halls, constantly watching for suspicious activity. They can spot someone trying to break into a room or tampering with the fire alarm. SIEM is like the security control room, where all the information from the CCTV, motion sensors, and alarms comes together. The security team in the control room can see the big picture and identify patterns that might indicate a larger threat. For example, they might notice that someone has been accessing restricted areas after hours, combined with unusual network activity. This could suggest a potential insider threat or a compromised account.

Now, imagine that the security team has a new member: a robot guard that can analyse vast amounts of data and identify patterns and anomalies that

The Security Team (cont'd)

humans might miss. It can also predict potential threats based on past events and current trends. The robot works alongside the human security team, providing valuable insights and helping to keep the school safe.

Like this robot guard, AI can enhance network monitoring by providing advanced analytics, automation, and threat intelligence. However, it's important to remember that AI is just a tool. It's still crucial to have human oversight and expertise to interpret the data and make informed decisions.

Like a real security team, network monitoring tools work together to protect your digital assets. IDPS provides real-time monitoring and prevention; SIEM offers an overview of network security. Together, they create a robust defence-in-depth strategy that helps keep the school's network safe and secure.

8

FINDING THE CRACKS BEFORE THEY CRUMBLE

Vulnerability Management

Just as a building requires regular inspections and maintenance to prevent structural issues, a school's network and systems need constant scrutiny to identify and address vulnerabilities before they can be exploited by cybercriminals.

This chapter explores the critical aspects of vulnerability management, emphasising the importance of regular security audits, penetration testing, and risk-based remediation. Think of vulnerability management as a proactive health check for your school's IT infrastructure, identifying potential weaknesses before they become major problems, like finding a loose tile on the roof before it falls and hits someone on the head.

Regular Security Audits: The Comprehensive Check-up

Regular security audits are like annual check-ups for your school's digital health. They provide a comprehensive assessment of the security posture, identifying potential vulnerabilities across the network, systems, and applications. Unlike relying on a single compliance report, regular audits offer a continuous view of security, allowing schools to track progress and identify emerging risks. External auditors can provide an unbiased perspective, bringing specialised expertise to identify vulnerabilities that might be missed by internal teams. These audits can uncover a wide range of issues, from outdated software and weak passwords to misconfigured firewalls and insecure access controls. The audit report should not just list vulnerabilities but also offer actionable recommendations for improvement.

Regular audits should cover various aspects of the school's IT infrastructure, including:

- **Network Infrastructure:** Reviewing firewall rules, router configurations, and network segmentation.
- **Server Security:** Assessing server hardening, patch management, and access controls.
- **Application Security:** Identifying vulnerabilities in web applications and other software.
- **Data Security:** Evaluating data encryption, access controls, and data loss prevention measures.
- **Physical Security:** Assessing physical access to server rooms and other critical infrastructure.
- **User Access and Authentication:** Reviewing password policies, user privileges, and multi-factor authentication.

Penetration Testing: The Simulated Attack

Penetration testing, sometimes referred to as "ethical hacking," takes vulnerability management a step further by simulating real-world attacks. Engaging qualified professionals to conduct penetration testing is like hiring a security expert to try and break into your school's network. They use the same tools and techniques as malicious hackers to identify system weaknesses and exploit vulnerabilities. This process helps to uncover vulnerabilities that might not be detected by automated scans or security audits.

Penetration testing can involve various techniques, including:

- **External Penetration Testing:** Simulating attacks from outside the school network, attempting to gain access to internal systems.
- **Internal Penetration Testing:** Simulating attacks from within the school network, testing the security of internal systems and data.
- **Web Application Penetration Testing:** Focusing on identifying vulnerabilities in web applications, such as SQL injection and cross-site scripting.
- **Social Engineering Testing:** Simulating phishing attacks and other social engineering techniques to test user awareness and susceptibility to manipulation.

Penetration tests provide valuable insights into the school's security posture and highlight areas that need immediate attention. The penetration testing report should include detailed information about the identified vulnerabilities, the potential impact of exploitation, and recommendations for remediation.

The Case of the Copycat Website

One morning, I received a panicked call: "Our website has been hacked! Someone has replaced the school logo with a picture of a dancing banana!" I rushed to my computer, heart pounding, expecting to find a scene of digital carnage. But the website looked perfectly normal. No dancing bananas, no defaced content, nothing out of order. Confused, I contacted our web hosting provider, who confirmed that there had been no unauthorised access to the server. It was then that I realised the truth: some tech-savvy students had created a near-perfect replica of the school website, complete with the dancing banana logo, and had cleverly redirected some internal links to their prank site. It was harmless, but it highlighted a potential vulnerability in our system. We quickly tracked down the culprits (who, to their credit, had some impressive coding skills), and I learned a valuable lesson about the importance of penetration testing. By simulating real-world attacks, it can uncover vulnerabilities that might otherwise go unnoticed, saving you from potential embarrassment (and a whole lot of dancing bananas).

Remediation: Fixing the Cracks

Once vulnerabilities have been identified through security audits and penetration testing, the next step is remediation. This involves taking corrective action to address the identified weaknesses and reduce the risk of exploitation. Prioritising vulnerabilities based on risk is essential. Not all vulnerabilities are created equal. Some vulnerabilities pose a greater threat than others, depending on the severity of the vulnerability and the potential impact of exploitation. A penetration test report would usually include a risk assessment section, outlining the likelihood and impact of each vulnerability found so you can prioritise remediation efforts accordingly.

Remediation can involve various actions, such as:

- **Patching Software:** Updating software and operating systems to address known vulnerabilities.
- **Configuring Systems Securely:** Changing system settings to improve security.
- **Implementing Access Controls:** Restricting access to sensitive data and systems.
- **Strengthening Passwords:** Enforcing strong password policies and implementing multi-factor authentication.
- **Training Users:** Educating users about security best practices and how to identify and report suspicious activity.

Remediation requires continuous monitoring and follow-up. Once a vulnerability has been remediated, it should be retested to ensure that the fix is effective.

Building a Robust Vulnerability Management Programme

Implementing an effective vulnerability management programme requires a comprehensive approach that includes:

- **Regular Assessments:** Conducting regular security audits and penetration testing.
- **Risk-Based Prioritisation:** Prioritising remediation efforts based on risk.
- **Timely Remediation:** Addressing vulnerabilities promptly.
- **Continuous Monitoring:** Continuously monitoring systems for new vulnerabilities.
- **Collaboration:** Working with IT staff, vendors, and other stakeholders to remediate vulnerabilities.

By implementing a robust vulnerability management programme, schools can significantly reduce their risk of cyberattacks and protect their valuable data and systems. It is an investment in the long-term security and stability of the school's digital environment, ensuring that the foundation upon which education thrives remains strong and resilient. Just as a well-maintained building stands strong against the elements, a well-managed digital infrastructure stands strong against cyber threats.

The Building Inspector

Think about your school building. Regular security audits are like having a building inspector come in to check the structure for any potential problems. They might find cracks in the foundation, leaky pipes, or faulty wiring.

Penetration testing is like having a team of engineers try to find weaknesses in the building's design. They might try to find ways to bypass security systems or exploit structural flaws. Remediation is like fixing those problems. It's about patching the cracks, fixing the leaks, and upgrading the wiring.

Just as with a real building, your school's IT infrastructure should have regular inspections and maintenance to remain in good shape. Vulnerability management is the process of finding and fixing those weaknesses before they can cause serious problems. It's about being proactive and taking steps to protect your school's digital assets.

9

EXTENDING THE CIRCLE OF TRUST

Third-Party Risk Management

In today's interconnected world, schools rarely operate in isolation. They rely on a network of third-party vendors for various services, from cloud storage and pupil information systems to catering services and transportation. While these partnerships are essential for efficient operations, they also introduce potential security risks. Each third-party vendor represents an extension of the school's digital perimeter, and a vulnerability in their systems can become a vulnerability for the school itself.

This chapter explores the crucial aspects of third-party risk management (TPRM), emphasising vendor assessments, contractual obligations, and ongoing monitoring. Think of your school's data as a precious family heirloom. You wouldn't just hand it to anyone, would you? Third-party risk management is about carefully vetting anyone who has access to that heirloom, ensuring they will handle it with the same care you would.

Vendor Assessments: Vetting the Partners

Before entrusting a third-party vendor with access to your school's systems or data, a thorough security assessment should be conducted. This assessment should evaluate the vendor's security posture, identify potential vulnerabilities, and determine their ability to protect sensitive information. It's like doing a DBS check before hiring a new member of staff at your school. You want to make sure they are trustworthy and capable.

Vendor assessments can involve various methods, including:

- **Questionnaires:** Sending vendors detailed questionnaires about their security practices, policies, and controls. These questionnaires should cover areas such as data encryption, access control, incident response, and vulnerability management.
- **Security Audits:** Conducting security audits of the vendor's systems and facilities. This can involve reviewing security logs, examining physical security measures, and testing network security.
- **Reviewing Certifications:** Checking if the vendor holds relevant security certifications, such as ISO 27001 or Cyber Essentials Plus. While certifications are not a guarantee of security, they can provide some assurance of the vendor's commitment to security best practices.
- **Background Checks:** For vendors who have access to highly sensitive data, such as pupil records, DBS checks on key staff may be necessary.

The scope and depth of the vendor assessment should be proportionate to the sensitivity of the data and the level of access granted to the vendor. For vendors who handle highly sensitive data, a more rigorous assessment, including on-site audits and penetration testing, may be necessary.

Contractual Obligations: Setting the Ground Rules

Once a vendor has been selected, it is crucial to include robust data sharing agreements and cybersecurity requirements in the contract. These contractual obligations serve as a legally binding agreement, outlining the responsibilities of both the school and the vendor regarding data security and privacy. It's like setting clear rules for a babysitter before leaving your children in their care. You want to make sure they understand what they are responsible for.

Key contractual provisions should include:

- **Data Ownership and Usage:** Clearly define who owns the data and how it can be used. Specify that the vendor can only use the data for the purposes outlined in the contract and cannot share it with other parties without consent.
- **Data Security Requirements:** Specify the security controls that the vendor must implement to protect the data, such as encryption, access control, and vulnerability management. Require the vendor to comply with relevant security standards and regulations.
- **Incident Response:** Outline the procedures for reporting and responding to security incidents. Require the vendor to notify the school immediately of any security breaches.
- **Liability:** Clearly define the liability of each party in the event of a data breach or other security incident.
- **Data Breach Notification:** Include provisions for notifying affected individuals and regulatory authorities in the event of a data breach, in compliance with relevant regulations.
- **Right to Audit:** Include a clause that grants the school the right to audit the vendor's security practices and controls.

Ongoing Monitoring: Keeping a Constant Watch

Even after a vendor has been assessed and a contract has been signed, ongoing monitoring of their security posture is essential. The threat landscape is constantly evolving, and a vendor's security practices can change over time. Regular monitoring helps to identify any emerging risks and ensure that the vendor is continuing to meet the school's security requirements. It's like checking in on the babysitter periodically, to see if everything is going smoothly.

Ongoing monitoring can involve various methods, including:

- **Regular Security Reviews:** Conducting periodic reviews of the vendor's security logs, vulnerability scans, and incident response plans.
- **News and Alerts:** Staying informed about any security incidents or vulnerabilities reported about the vendor.
- **Third-Party Risk Management Platforms:** Utilising third-party risk management platforms that provide continuous monitoring of vendor security posture.
- **Communication:** Maintaining open communication with the vendor's security team to discuss any security concerns or updates.

The Bare Necessities

My school was considering a new online platform but, before we signed on the dotted line, I sent the vendor a detailed cybersecurity questionnaire. Their response was... well, a little unexpected. They assured me that they took cybersecurity very seriously, but their answers were vague and evasive. Then, to top it off, their final email ended with the phrase "please **bare** with us as we improve our security posture." I couldn't resist. I replied with a cheeky email saying, "I appreciate your commitment to cybersecurity, but what does nudity have to do with it?!" To their credit, they saw the humour and responded by fully covering their procedures. But this highlighted the importance of clear communication with vendors, especially when it comes to complex topics like cybersecurity. It's like ordering a coffee – you wouldn't just say "I want a coffee," would you? You'd specify the size, the type of milk, and whether you want it with sugar or not. The same goes for cybersecurity. Be specific, ask questions, and don't be afraid to go beyond the bare necessities!

Building a Robust Third-Party Risk Management Programme

Implementing an effective TPRM programme requires a comprehensive and ongoing effort. Schools should:

- **Develop a TPRM Policy:** Establish a clear policy that outlines the school's approach to managing third-party risks.
- **Identify Critical Vendors:** Identify the vendors who have access to the most sensitive data and prioritise their assessments and monitoring.
- **Establish a Process:** Develop a standardised process for conducting vendor assessments, negotiating contracts, and monitoring vendor security.
- **Use Technology:** Utilise third-party risk management platforms and other tools to automate and streamline the TPRM process.
- **Train Staff:** Train staff on the importance of TPRM and their role in the process.

Third-party risk management helps schools significantly reduce their exposure to security threats and ensure the continued protection of their valuable data. It is an essential element of a comprehensive cybersecurity strategy, recognising that security is only as strong as its weakest link. Just as a responsible parent carefully chooses their child's caregivers, a responsible school must carefully select and monitor its third-party vendors.

Your Trusted Babysitters

Think of your school's data as your own children. Third-party vendors are like the babysitters you hire to look after them. Vendor assessments are like interviewing the babysitters, checking their references and DBS status, and making sure they are qualified. Contractual obligations are like setting clear rules for the babysitters, telling them what they are allowed to do and what they are not. Ongoing monitoring is like checking in on the babysitters periodically to see if everything is going well, and the children are safe.

Just like you would carefully vet and monitor anyone who cares for your children, you should take the same approach with third-party vendors who have access to your school's data. A robust TPRM programme can extend the circle of trust and ensure that your data is protected even when it is in the hands of others.

10

STRENGTH IN NUMBERS

Collaboration & Information Sharing

Schools are facing a constant barrage of new cyber threats and vulnerabilities. No single school, no matter how well-resourced, can effectively combat these challenges in isolation. Collaboration and information sharing are essential for staying ahead of the curve, pooling resources, and leveraging collective knowledge.

This chapter explores the crucial role of collaboration and information sharing in strengthening school cybersecurity, emphasising the importance of exchanging threat intelligence with other institutions and partnering with external experts. Think of cybersecurity as a community garden. Individual gardeners can tend their own plots, but sharing knowledge and working together makes the entire garden thrive.

The Plagiarised PowerPoint (or, Why Sharing is Caring)

I'll never forget the time Mr. Borrowed Ideas, a history pupil with a penchant for, shall we say, "repurposing" online resources, inadvertently sparked a school-wide panic. He had downloaded a seemingly innocuous PowerPoint presentation from a not-so-reputable website for his upcoming assignment on the Roman Empire. Unbeknownst to him, the presentation was laced with a particularly nasty strain of malware. Within minutes of opening the file, Mr. Borrowed Ideas' laptop was flashing error messages faster than a Roman chariot race. The malware quickly spread to other devices on the network, causing widespread disruption and a flurry of frantic calls to the IT helpdesk. Thankfully, I had recently participated in an online school IT Directors' forum where we had discussed shared threat intelligence and best practices. One of the other, more

The Plagiarised PowerPoint (cont'd)

well-resourced schools had already encountered this exact malware and had developed a handy tool to neutralise it. We quickly deployed the tool, quarantined the infected devices, and restored order to the digital empire. Mr. Borrowed Ideas learned a valuable lesson about the dangers of "borrowing" from untrusted sources, and I gained a newfound appreciation for the power of collaboration and information sharing in cybersecurity.

It's like sharing your secret recipe for the best chocolate chip cookies – you might be hesitant at first, but the joy of sharing (and the collective benefit of delicious cookies) far outweighs the risk of someone stealing your culinary thunder.

Information Sharing: The Power of Collective Intelligence

By exchanging threat intelligence with other schools, organisations, and industry groups, schools can stay informed about the latest threats, vulnerabilities, and attack techniques. This shared knowledge empowers schools to proactively defend against emerging threats and avoid falling victim to attacks that have already targeted others. It's like a neighbourhood watch programme where everyone shares information about suspicious activity, making the entire neighbourhood safer.

Information sharing can take many forms, including:

- **Participation in Information Sharing and Analysis Centres (ISACs):** ISACs are industry-specific groups that facilitate the sharing of cybersecurity information among members. Schools can join education-focused ISACs to receive timely alerts about emerging threats and share their own experiences.
- **Collaboration with Peers:** Networking with IT colleagues at other schools allows for informal information sharing and collaboration. This can involve sharing best practices, discussing security challenges, and coordinating incident response efforts.
- **Engagement with Government Agencies:** Government agencies, such as the National Cyber Security Centre (NCSC), often provide threat intelligence and cybersecurity resources to schools. Schools should actively engage with these agencies to stay informed about the latest threats.
- **Use of Threat Intelligence Platforms:** Threat intelligence

platforms, often hosted by your anti-malware provider, aggregate and analyse threat data from various sources, providing schools with actionable insights about emerging threats.

- **Open-Source Intelligence (OSINT):** Leveraging publicly available information, such as security blogs, vulnerability databases, and research papers, to stay informed about the latest threats and vulnerabilities.

Effective information sharing requires trust and reciprocity. Schools must be willing to share their own experiences and insights to benefit from the collective intelligence of the group. It is a two-way street; you get what you give.

Collaboration: Pooling Resources & Expertise

Schools rarely have the resources or expertise to handle all aspects of cybersecurity in-house. Collaborating with other schools, industry partners, and external cybersecurity experts can provide access to specialised knowledge, tools, and resources. This collaborative approach allows schools to leverage the expertise of others, fill gaps in their own capabilities, and improve their overall security posture. It's like forming a team of specialists to tackle a complex project. Each member brings their unique skills and expertise to the table, making the team stronger than any individual.

Collaboration can involve various forms, including:

- **Joint Procurement:** Schools can collaborate to purchase cybersecurity products and services at discounted rates, leveraging their collective buying power.
- **Shared Services:** Schools can share cybersecurity services, such as incident response, vulnerability scanning, and

security awareness training, reducing costs and improving efficiency.

- **Partnerships with Cybersecurity Companies:** Schools can partner with cybersecurity consultants to gain access to specialised expertise and tools. They can assist with everything from risk assessments and penetration testing to incident response and security awareness training.
- **Participation in Cybersecurity Consortia:** Cybersecurity consortia bring together schools, industry partners, and government agencies to collaborate on cybersecurity initiatives, share best practices, and develop joint solutions.

Building a Culture of Collaboration & Information Sharing

Creating a culture of collaboration and information sharing requires a concerted effort. Schools should:

- **Foster Trust:** Build trust with other schools and organisations to encourage open communication and information sharing.
- **Establish Communication Channels:** Establish clear communication channels for sharing information, such as email lists, online forums, and regular meetings.
- **Participate in Communities:** Actively participate in IT communities and forums to connect with other professionals and share knowledge.
- **Encourage Collaboration:** Encourage staff to collaborate with colleagues at other schools and participate in joint initiatives.
- **Recognise and Reward:** Recognise and reward staff for their contributions to information sharing and collaboration.

By embracing collaboration and information sharing, schools can significantly strengthen their cybersecurity defences. It's about recognising that by working together, we can achieve more than we ever could alone. Just as a community garden thrives on shared knowledge and effort, so too does the cybersecurity community.

Avengers Assemble!

Imagine your school's IT team as a superhero. They are strong and capable, but they can't do everything on their own.

Information sharing is like having a network of other superheroes who exchange information about the latest villains and their plans.

Collaboration is like forming a team of superheroes, each with their own special powers, to fight a common enemy. Together, they are much stronger than any individual hero.

Like the Avengers, schools should work together to combat ever-evolving cyber threats. Information sharing provides the intelligence, and collaboration provides the muscle.

11

FROM COMPLIANCE TO VIGILANCE

Cultivating a Culture of Cybersecurity

Cybersecurity is not a destination; it's a journey. It is not a checklist to be completed or a box to be ticked. It is an ongoing process that demands constant vigilance, continuous education, and practical, adaptable measures. While compliance with regulations and standards is important, it is merely a starting point. True cybersecurity lies in fostering a culture where security is not just a technical concern but a shared responsibility, woven into the fabric of the school community.

This concluding chapter emphasises the importance of moving beyond mere compliance and cultivating a proactive, hands-on approach to cybersecurity, where everyone understands their role in safeguarding the school's digital assets and nurturing a safe and secure learning environment.

Beyond Compliance: Embracing a Proactive Stance

Compliance provides a baseline of security practices, but it doesn't guarantee complete protection. Cybercriminals are constantly adapting their tactics, and simply adhering to a set of rules is not enough to stay ahead of the curve. Schools must adopt a proactive approach, anticipating potential threats and adapting their defences accordingly. This means moving beyond a reactive mindset, where security is only addressed after an incident occurs, and embracing a proactive stance, where security is integrated into every aspect of the school's operations.

A proactive approach to cybersecurity involves:

- **Continuous Education:** Regularly educating staff, pupils, and parents about cybersecurity best practices, emerging threats, and the school's security policies. This education should be engaging, relevant, and tailored to the specific needs of each group.
- **Regular Updates and Patching:** Keeping software and systems up to date with the latest security patches is crucial for closing vulnerabilities and preventing exploitation. Automating updates whenever possible can help ensure that systems are always protected.
- **Vulnerability Management:** Regularly scanning for vulnerabilities and remediating them promptly. This includes conducting security audits, penetration testing, and vulnerability assessments.
- **Incident Response Planning:** Developing and regularly testing an incident response plan to ensure that the school is prepared to handle a cyberattack or other security incident. This plan should outline roles and responsibilities, communication strategies, and technical procedures for containing, eradicating, and recovering from an incident.
- **Security Awareness Training:** Conducting regular security awareness training to educate staff and pupils about common threats, such as phishing attacks and social engineering, and how to recognise and report suspicious activity.

Building a Culture of Cybersecurity: A Shared Responsibility

A strong cybersecurity posture relies not just on technology and policies, but also on the human element. Creating a culture of cybersecurity means making security a shared responsibility, where everyone understands their role in protecting the school's digital assets. This requires fostering a sense of ownership and accountability, empowering individuals to take an active role in security.

Building a culture of cybersecurity involves:

- **Leadership Commitment:** Demonstrating strong leadership commitment to cybersecurity. School leaders should champion security initiatives, allocate resources, and emphasise the importance of security.
- **Empowerment & Accountability:** Empowering individuals to take responsibility for security and holding them accountable for their actions. This includes providing clear guidelines and expectations, as well as consequences for security violations.
- **Open Communication:** Fostering open communication about security issues. Encourage staff and pupils to report suspicious activity without fear of reprisal.
- **Continuous Improvement:** Regularly reviewing and updating security policies and procedures to reflect changes in the threat landscape and best practices. Continuously seeking ways to improve the school's security posture.
- **Integration with School Values:** Integrating cybersecurity into the school's overall values and mission. This reinforces its importance and makes it a part of the school's identity.

The Path Forward: A Continuous Journey

Cybersecurity is not a one-time fix. As technology evolves and cyber threats become more sophisticated, schools must adapt their strategies and practices to stay ahead of the curve. This requires a commitment to continuous learning, collaboration, and improvement. By embracing a proactive, hands-on approach to cybersecurity and cultivating a culture of shared responsibility, schools can create a safer and more secure learning environment for pupils and staff. It is about recognising that cybersecurity is not just a technical issue; it is a human issue, and it requires a collective effort to protect our digital future. Just as we teach our children to look both ways before crossing the street, we must also teach them to be responsible digital citizens, aware of the risks and empowered to protect themselves and their community. The journey towards a truly secure digital future begins with cultivating a culture of cybersecurity, where vigilance, education, and proactive measures are not just best practices, but ingrained habits.

A Digital Immune System

Your school's cybersecurity is like its digital immune system. Compliance is like having a healthy diet and getting regular exercise – it strengthens the body's natural defences. But a proactive approach is like getting vaccinated – it provides targeted protection against specific threats. Building a culture of cybersecurity is like having a strong community of white blood cells, constantly vigilant and ready to fight off any infection. Each individual cell plays a crucial role in protecting the body, just as each member of the school community plays a crucial role in protecting the school's digital assets.

Just like a healthy immune system requires constant care and attention, a strong cybersecurity posture requires ongoing effort and commitment. It's not enough to simply react to threats; we must proactively build our defences and cultivate a culture of vigilance.

ACKNOWLEDGEMENTS

Writing this book has been an enlightening and rewarding journey, made possible by the support, guidance, and encouragement of numerous individuals.

I owe a huge thank you to my dear friends, **Katie and Alex Cunningham**, whose expertise and insightful feedback have greatly enhanced the quality of this book. It was Katie who first suggested in 2023 that a short paper I had written on cybersecurity might be turned into a book. She is a tireless mum of two, an IT consultant, and a Chelsea FC fan (nobody's perfect); her husband Alex is an English teacher and self-confessed Luddite but asked many questions that challenged me to think about how I presented the ideas in this book. Your encouragement and good humour were invaluable throughout this process.

The spark for this book likely ignited in January 2020, when I was invited by **Ian Phillips** to speak at EdTech UK Conversations, co-hosted by the Independent Schools Council (ISC) Digital Advisory Group. I have always felt more confident expressing my thoughts on paper than in front of a room full of people, so when, to my surprise, everyone was still wide awake at the end, I started to consider that perhaps I did have something worthwhile to share about my work and experiences. The notes from my presentation on the importance of stakeholder relationships in delivering successful digital transformation projects proved useful in drafting some of Chapters 9 and 10 of this book. Ian has remained a valuable and respected sounding board, as has another colleague who spoke at that event, **Gary Henderson**, now Vice-Chair of that ISC group.

For Chapter 3, I drew on my notes and title – A Firm Foundation – from a 2021 talk to fellow IT professionals and school leaders on the importance of getting a school's infrastructure right. My analogy of building roads before Ferraris prompted a few chuckles that day. That event was hosted by **John Sainthouse** and **Nick Donoghue**, with whom I have worked closely on several projects, including cybersecurity, at different schools over the years. Both remain good friends and consultants. The presentation on cybersecurity that day came from **Anna Kempster**, who is part of a close network of IT colleagues with whom I keep in contact. Anna hosts an online forum where we all get together regularly to exchange information and share ideas – exactly the sort of thing I advocate for in Chapter 10.

That ever-expanding network includes other trusted friends and colleagues like **Jeremy Davies, Tony Whelton, Alex Henderson, Jessen Chen, Dan Raymond**, **Laura Knight**, **Christopher O'Mahony** and **Tony Phillips**, among others. I have learnt a lot from all of you during our many discussions and informal get-togethers, and had a lot of fun along the way. Yes, we IT folks do know how to have a good time!

I always imagined that my first book would be a novel. But then, never in my wildest dreams did I imagine having a career in technology. IT was meant to be just a short diversion while on sabbatical from a then fairly successful journalism career. Yet, here I am 26 years later! It wouldn't have been possible without people like **Adeola Gbadamosi** and the late **Salim Hariff** who put their faith in me early on and gave me a solid foundation of core skills on which my IT career has been built. I have also been very fortunate to have had some great managers and mentors (and the occasional tor-mentor) to guide me along the way – **Mark Mackenzie Crooks**, who provided a much-needed shot in the arm at a time when I was at risk of stagnating; **Geoff Wilmot**, one of the kindest and most inspiring individuals I have ever had the privilege to work with; **Bob Ukiah, Michael Michaelides, Kenneth Evans, Terry Stevens** and **Nigel Slater.** I am grateful to all of them for their contributions to my personal and professional development.

Over the past 26 years I have also had the pleasure of working alongside some immensely talented, hard-working and innovative IT colleagues – among them **Fay, Rahim, Lianna, Paul, Sujata, Steve, Michael, Mary, Bright, Kuldev, Amit, Sevcan, Myra, Colin, Anthony** and **Yanai,** who now runs his own cybersecurity consultancy. You have been my fellow soldiers in the IT trenches, with whom I have laughed, squabbled, and even cried. But we haven't done it alone: we were regularly supported, encouraged, challenged and, yes, at times frustrated, by school heads, senior leaders, teachers, admin professionals, pupils, parents and governors, who make up those important school communities I highlight in Chapter 1.

If some of the characters in this book – Miss Stern, Mr. Grumbles, Mr. Borrowed Ideas, Mr. Click-Happy, and Mrs. Panicked – bear any resemblance to actual persons, it is because they are. Their true identities will remain secret, but I know some of them are quite thrilled to have been immortalised in a book! You know who you are. Thank you for making my days even more interesting.

Almost no one, including close friends and family, knew about this little project until it was almost done. That's just how I prefer to do things – stay calm and get on with it. But I am grateful to Mum and Dad, my siblings, and my friends, who celebrated and cheered me on once you had got over my surprise announcement that I had written a book. Your encouragement and understanding have been a source of strength and motivation.

As I wrote during the quiet hours, usually late into the night after a full day's work, I thought often of my younger brother, **Joel,** who passed away in 2022. He would have been a great champion of my efforts, and how I wish he was still here to read the final product. He would have surely found something in this book he could use to tease me mercilessly for years to come.

Finally, I extend my gratitude to the readers of this book. It is my hope that the insights and strategies shared here will empower you to enhance cybersecurity in your schools, ensuring a safer digital environment for pupils and staff.

Thank you all for being a part of this journey and for your unwavering support.

Sincerely,

David

[1]   https://www.gov.uk/government/statistics/cyber-security-breaches-survey-2024/cyber-security-breaches-survey-2024-education-institutions-annex

[2] https://www.infosecurity-magazine.com/news/schools-hit-by-cyberattacks-in/

[3]   https://www.microsoft.com/en-us/security/blog/2024/10/10/cyber-signals-issue-8-education-under-siege-how-cybercriminals-target-our-schools/

[4]   https://www.highereddive.com/news/moodys-rates-education-sector-high-cyber-risk-2024/733462/

[5] https://www.ncsc.gov.uk/section/education-skills/cyber-security-schools

About the Author

David Nanton brings a unique blend of experience to the world of cybersecurity. A former newspaper journalist with a nose for a good story, he transitioned to the IT sector over 25 years ago, dedicating most of his career to supporting schools in the UK. His experience spans both the State and independent sectors, giving him a broad perspective on the challenges and triumphs of educational technology. David's passion for cybersecurity stems from his first-hand experience of the growing threats faced by schools and his desire to empower educators with the knowledge and tools they need to create a safe and secure learning environment.